SpringerBriefs in Applied Sciences and Technology

SpringerBriefs present concise summaries of cutting-edge research and practical applications across a wide spectrum of fields. Featuring compact volumes of 50 to 125 pages, the series covers a range of content from professional to academic.

Typical publications can be:

- A timely report of state-of-the art methods
- An introduction to or a manual for the application of mathematical or computer techniques
- A bridge between new research results, as published in journal articles
- A snapshot of a hot or emerging topic
- An in-depth case study
- A presentation of core concepts that students must understand in order to make independent contributions

SpringerBriefs are characterized by fast, global electronic dissemination, standard publishing contracts, standardized manuscript preparation and formatting guidelines, and expedited production schedules.

On the one hand, **SpringerBriefs in Applied Sciences and Technology** are devoted to the publication of fundamentals and applications within the different classical engineering disciplines as well as in interdisciplinary fields that recently emerged between these areas. On the other hand, as the boundary separating fundamental research and applied technology is more and more dissolving, this series is particularly open to trans-disciplinary topics between fundamental science and engineering.

Indexed by EI-Compendex, SCOPUS and Springerlink.

More information about this series at https://link.springer.com/bookseries/8884

Introduction

The public studies show that the abrasive water jet has become one of the widely researched topics due to its many useful applications in manufacturing field. The USA is the most influential and active country in the abrasive water jet machining research (Anwar, 2019).

Water Jet Machining is one of the most progressive machining technologies. Owing to the current high demands on production quality and productivity, it can be considered as a progressive CNC controlled technology. The modern era calls for the continuous research results in this area (Radvanská, 2010; Hloch, 2008). Nowadays, the abrasive water jet technology is much in use for cutting materials into complex shapes with high dimensional accuracy (Kumar, 2016).

The topic of the submitted monograph was selected with regard to the high topicality and use of the Water Jet Machining technology in engineering practice, while focusing on the Abrasive Water Jet Cutting technology as one of the most widely utilized applications of water jet machining. Theoretical part of the monograph describes the AWJC technology in more detail. Its important subchapter deals with the process of piercing the materials, which is addressed in the monograph. Piercing of materials is presented from the aspect of the current methods of its implementation used in practice. We can state that, at present, piercing of material is performed for all types and thicknesses of materials, always within the residual material.

The experimental part of the monograph submits a proposal for an innovative piercing performance, along with experimental verification of the monograph's objective—to perform piercing of material on the workpiece contour when cutting by AWJC of material DIN 1.4301.

References

Anwar, S. et al. (2019). Bibliometric analysis of abrasive water jet machining research. *Journal of King Saud University—Engineering Sciences, 31*(3), 262–270.
Hloch, S., Valíček, J. (2008). *Vplyv faktorov na topografiu povrchov vytvorených hydroabrazívnym delením.* Prešov: FVT TU (p. 125). ISBN: 978-80-553-0091-7.

Kumar, R. et al. (2016). Surface Integrity Analysis of Abrasive Water Jet-Cut Surfaces of Friction Stir Welded Joints. *International Journal of Advanced Manufacturing Technology, 88,* 1687–1701.

Radvanská, A. (2010). Abrasive waterjet cutting technology risk assessment by means of failure modes and effects analysis method. In: *Technical Gazette* (p. 121–128), Vol. 17(1), ISSN 1330-3651.

Contents

About the Authors

Ivana Kleinedlerová Graduated from STU MTF in Trnava, the study program of Production Technologies. Her dissertation thesis dealt with water jet cutting.

Works for STU MTF in Trnava, teaching Computer-Aided Production Technologies, Fundamentals of Design and Technical Documentation, Fundamentals of Economics and Management and Enterprise Economy. Her professional, scientific, and publication activities are mainly focused on the field of Production Engineering, Water Jet Cutting in particular.

Peter Kleinedler Graduated from STU MTF in Trnava, the study program of Engineering Technologies and Materials. His doctoral thesis was focused on Metrology.

Currently works for STU MTF in Trnava, teaching the subject of Fundamentals of Production Technologies, and simultaneously managing the MTF STU Training Center in Dubnica nad Váhom, being its Head since 2010. His professional and publication activities are focused on the field of Mechanical Engineering.

Abbreviations

A	Abrasive
a_1, b_1, c_1, a_2, b_2	Final values of the approximated line at the 2nd degree polynomial substitution
ACJ	Abrasive Cryogenic Jet
AWJC	Abrasive Water Jet Cutting
AWJM	Abrasive Water Jet Machining
b [mm]	Thickness of machined material
DIN 1.4301	Stainless steel
d_{z1} [mm]	Diameter of orifice
d_{z2}	Focusing nozzle
FAWJ	Fine Abrasive Water Jet
FAWJM	Fine Abrasive Water Jet Machining
Interpol	Mathematical approximation program,
$k.x, q$	Coefficients of linear approximation
L [mm]	Standoff distance
L_p [mm]	Length of the jet trajectory performed during piercing, length of piercing
M_a [g.min^{-1}]	Mass flow of abrasive
MESH	Abrasive grain size
N	Number of the jet transitions along the same trajectory
p [MPa]	Working pressure of liquid
S [mm^2]	Surface influenced by a jet in AWJC,
v_f [mm.min^{-1}]	A jet feed rate along the machined surface of the cut material, working rate
WJC	Water Jet Cutting
WJM	Water Jet Machining
α [°]	Angle of the jet deflection from its original (perpendicular) shape
β [°]	Angle of the jet impact on the material being machined

Chapter 1
Cutting by AWJC

Nowadays, advanced machining techniques are widely used for solving various issues in manufacturing operations that include machining high strength materials, production of complex shaped profiles, better surface features, capable of high levels of precision, miniaturization, reduction of waste and secondary operations, and lower production time. Among the various advanced machining techniques, AWJM has received more attention from researchers and practicing engineers in manufacturing industries due to its capability of extensive operations and excellent quality of the cutting edge obtained during this process much superior to others, as reported by previous researchers (Natarajan, 2020). Utilization of advanced materials is conditioned by their machinability. Conventional machining methods are frequently unsuitable for these purposes owing to deformation processes, thermal influence, low machining speed, and so on. The continuous trend of introducing new construction materials has also affected the development and industrial use of the new, unconventional machining technologies, including AWJM. Abrasive water jet is one of the few tools able to satisfy the requirements of technologists and adapt to the trend of developing the engineering materials of specific properties (Hloch, 2008). It is a very suitable and a widely used technology for machining the difficult-to-machine materials. An important group of the materials suitable for water jet machining is, among others, composite materials. A great advantage of water jet machining is the machined material free from the mechanical, thermal, and chemical impact. From the aspects of energy, economy, and environment, it is one of the advanced technologies; besides, it can be automated. These features make the Water Jet Machining technology universal, which is also due to the wide range of the technological applications and machining methods. Currently, the water jet technology provides a wide scale of production processes used in technology practice (Krajný, 1998).

A proof of the constant advancement of the technology is the development of new water jet applications, such as (DIN 8580, 2003):

© The Author(s), under exclusive license to Springer Nature Switzerland AG 2022
I. Kleinedlerová and P. Kleinedler, *Piercing of Materials with Abrasive Water Jet*,
SpringerBriefs in Applied Sciences and Technology,
https://doi.org/10.1007/978-3-030-92130-9_1

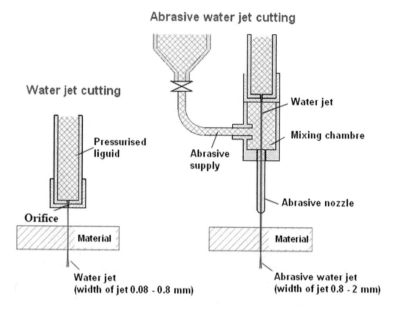

Fig. 1.1 A simplified model of differences between the WJC and AWJC technologies. *Source* Krajný (1998)

– Refining of materials: surface planning, polishing, roughening, peening.
– Machining of materials: cutting, drilling, turning, milling.
– Surface cleaning: cleaning, descaling, corrosion removal.

Currently, the most common water jet application is cutting. In terms of the working medium used, two basic cutting methods are distinguished (Vasilko, 1990):

- **WJC—Water Jet Cutting**,
- **AWJC—Abrasive Water Jet Cutting**.

Since authors of the scientific publications dealing with AWJC use different terminology, we chose the terminology according to the (VDI 2906, 1984) Standard and the following terms:

- Abrasive Water Jet Cutting,
- Focusing nozzle,
- v_f [mm min^{-1}]—feed rate of the jet movement along the machined surface of the material being cut, also called working speed,
- Residual material, a material of a semi-finished product, which remained after cutting the workpiece and can be used for further machining operations.

Figure 1.1 shows a model of the fundamental difference between Water Jet Cutting and Abrasive Water Jet Cutting.

Abrasive water jet shows several advantages, such as (Krajný, 1998):

- Application of high cutting speed, resulting in high accuracy of cut workpieces,
- Possibility to cut both, planar as well as complex surfaces within narrow tolerances,
- Minimal thermal impact on the surface, cold cutting process,
- Low deformation stress in the treated surface,
- Possibility to cut in different directions without loss of the jet efficiency,
- Possibility to control the jet trajectory by a computer,
- Relatively high energy efficiency (approx. 50%),
- Narrow cutting gap (0.8–2 mm).

Disadvantages of Abrasive Water Jet technology include (Krajný, 1998):

- High investment and operating costs,
- Relatively low jet feed rate on the machined surface of the material being cut, when cutting "hard" materials,
- Uneven, rough surface with, e.g. scratches and grooves,
- Very high noise level (more than 100 dB).

1.1 Principle of Cutting the Materials by AWJC Technology

Pulsating water jet is a hybrid process of ultrasonic machining and waterjet machining. The continuous jet of the water is induced with ultrasonic disturbances resulting in forced breakup of the jet (Nag, 2021). The principle of cutting the materials using abrasive water jet is based on the action of high kinetic energy of a water jet containing abrasive. At the point of the jet contact with the machined material, a mechanically abrasive removal of the machined material takes place owing to the action of an artificially induced and controlled erosion process of material cutting. Thus, AWJC consists in gradual removal of material, which is due to the mechanical impact of a narrow water jet containing abrasive, at a high speed and kinetic energy per an area unit. Abrasive, as an important additive, multiplies the mechanical effect of the jet impact on the material. Physical nature of cutting materials dwells in the fact that the jet reaches a speed two- to four-times higher than the speed of sound, behaving (in terms of its effect on the surface of the material being machined) as a solid body. During cutting, the material resists, as a result of which the jet gradually loses its kinetic energy and shape, and ultimately bends while deflecting from its original perpendicular shape (Fig. 1.2) (Kulekci, 2002).

This phenomenon is called a jet retardation. At each point of the material, the jet moves along a rounded trajectory in which the radius of curvature changes with the previous length of the arc (Kulekci, 2002). AWJC is a technology that allows cutting all types of materials in thicknesses from 0.1 to 350 mm, without damaging them by direct water contact (Vasilko, 1990). Cutting as well as quality of machined surface is significantly influenced by the jet retardation and its deflection to the side, which is given by the type and thickness of the material being machined. Mechanism of

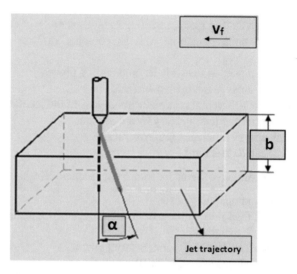

Fig. 1.2 Jet deflection in AWJC. *Source* Rezmat (2018)

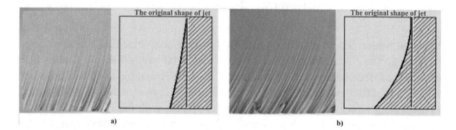

Fig. 1.3 Schematic representation of the jet and its deflection for the material **a** Jet deflection to the side when cutting material, thickness of 40 mm. **b** Jet deflection to the side with a significant manifestation of jet retardation associated with the formation of local non-cutting of the material, thickness of 80 mm

water jet cutting depends on the structure of the material being machined (Maňková, 2000).

Figure 1.3 shows the curve of the jet trajectory and its deflection for various thicknesses of the machined DIN 1.4301 material.

1.2   Course of AWJC

Cutting is a process in which a jet penetrates the entire thickness of the material, thereby dividing it into two separate parts. During cutting, a small amount of material is removed from the space of the cutting gap (Folkes, 2009).

The course of AWJC is as follows (Summers, 1997):

Phase 1—**Piercing**. After exiting the focusing nozzle and a very short passage through the environment, the jet enters into its first interaction with the material. The initial impact of the jet front part is followed by a continuous jet penetration into the material. At the point of impact, a formed opening gradually changes to a continuous hole; the size and shape of the hole are influenced by the piercing method used and the type and thickness of the material being machined. Piercing is associated with increased noise emission (125 dB) and spattering of the working mixture into the environment. The time interval of the jet action during its penetration into the machined material is expressed as the exposure time. It is the time interval required to form a continuous through hole in the material being machined. It is measured from the moment of the jet impacting the material surface up to its primary exit from the material being machined (Ohlsson, 1992).

Phase 2—**Start of cutting**. Formation of a continuous hole creates the preconditions for the natural removal of the working mixture, stabilizing simultaneously the jet shape. Subsequently, at a suitable speed, the entire volume of jet gradually penetrates further into the material (Summers, 1997).

Phase 3—**Cutting**. Continual movement of the jet cuts the entire thickness of material. While cutting, the jet deflects from the original perpendicular direction owing to the friction between the jet surface and the workpiece material (Hashish, 1988).

Phase 4—**End of cutting**. Exit of jet from the cut contour of the machined shape. The final cutting phase can be completed in several ways, depending on: material thickness, workpiece shape and size of residual material (Zeng & Kim, 1992).

AWJC can also be classified based on the sections of the process (Fig. 1.4) (Summers, 1997; Hloch, 2008).

The **entrance section of jet** is the beginning of the cutting gap formation. The section is characterized by various disruptions of materials. The jet reaches the maximum thickness of the cut, generally equal to the total thickness of the material being machined. In the **steady cutting section**, the cutting proceeds through a cyclic cutting process that continues until the jet reaches the end edge of the material. In this section, the full depth of the cut given by technological parameters is achieved.

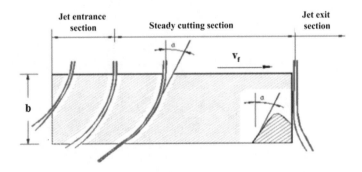

Fig. 1.4 Sections of AWJC. *Source* Hashish (1988)

The **exit section of jet** finishes the cutting process. In this section, the material is cut only to a certain depth. When the jet leaves the cut at the end of the material, leaving a triangular zone of uncut material, which proves that the cutting process is stable only to a certain depth (Summers, 1997; Hloch, 2008).

1.3 Technological Parameters of AWJC

The effect of abrasive water jet on the material being machined and the cutting efficiency both depend on the technological parameters used in cutting. Technological and operational characteristics of the abrasive water jet are quantified by the recommended ranges of parameters. Technological parameters are divided into groups based on various criteria. The parameters of major influence on the cutting efficiency are listed in Fig. 1.5 (Hassan, 2001).

Liquid pressure—p[MPa]

Industrially used devices for generating high pressure of liquid allow the use of working pressures with a continuously variable regulation in the range of 50–600 MPa. In general, the higher the pressure used, the greater the cutting efficiency (Zain, 2011). The high water pressure AWJM produces the better surface quality due to sufficient material removal and proper cleaning of debris from the machining zone as compared to the low water pressure and low abrasive mass ow rate (Mardi, 2021).

Fig. 1.5 The key technological parameters in AWJC. *Source* Krajný (1998)

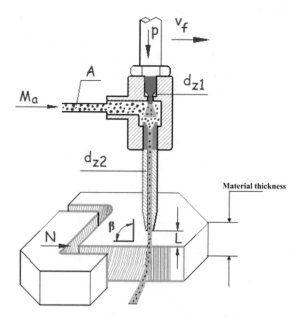

Morphology of the PWJ defines different erosion regimes based on the specific value of supply pressure and standoff distance (Hloch, 2020).

Jet feed rate on the machined surface of the material being cut—v_f [mm min^{-1}]

The feed rate of the jet on the machined surface of the material being cut is one of the decisive factors of the technology, influencing the achieved quality of the cutting surface and the shape of the cut workpieces. For individual types and thicknesses of materials, each company gradually develops its own cutting rates, along with the evaluation of the achieved surface quality and the shape of the cut workpieces (Zain, 2011).

Orifice (diameter)—dz_1 [mm]

The achieved depth of cut depends approximately linearly on the diameter of the water nozzle. The change of the inner diameter of orifice also affects the achieved quality and shape of the cut surface (Krajný, 1998).

Focusing nozzle (diameter, length, and geometry of the internal shape)—dz_2

The diameter and length of nozzle determine the properties of the exit jet (Vajdová, 2014). Constant flow of the high-erosion mixture brings about an undesired wear of the nozzle (Hloch, 2007), subsequently manifested as an enlarged inner diameter of the nozzle, which is reflected in the change of the shape and deterioration of the jet, as well as deterioration of the machined surface (Kleinedlerová, 2011; Kleinedlerová & Janáč, 2012). Size and shape of the focusing nozzle are one of the main factors affecting the quality and cutting performance (Vajdová, 2014).

Number of the jet transitions along the same trajectory—N

Increasing the number of jet transitions along the same trajectory without changing other technological parameters represents another possible way of achieving an increased thickness size of the material being cut (Hloch, 2008). A greater thickness of the cut can be achieved by multiple transitions of the jet along the same trajectory even at a constant pressure and the jet feed rate along the machined surface of the material being cut (Hashish, 1991).

Standoff distance—L [mm]

While passing through the environment between the focusing nozzle and the material, the originally coherent jet scatters and gradually disintegrated. The distance of the focusing nozzle mouth from the material surface is therefore as small as possible, i.e. 2–4 mm (Hlaváč, 1994).

However, the distance of the nozzle mouth from the surface of the machined material cannot be zero, which is due to (Hlaváč, 2000):

- possible contact of nozzle with unevenness of the workpiece, leading to potential damage,
- certain distance required for the expansion of the compressed mixture of water and abrasive.

The growing distance of the focusing nozzle mouth from the material surface leads to the reduced cutting speed and deterioration of the cutting surface quality (Hlaváč, 2000).

Abrasive (medium grain size, shape, density, and hardness)—*A*

The abrasive used for cutting should meet the following criteria: maximum hardness, affordability, suitable shape of abrasive grains, and health safety. The size of abrasive particles significantly affects the AWJC mechanism in terms of their entry into the water jet and the kinetic energy required to cut the material (Hlaváč, 1999). Abrasive particles of high roundness and low circularity are considered the best. The choice of a suitable abrasive material depends on the technical and economic possibilities, and, last but not least, on the environmental impact. In practice, the most widely used is the Garnet abrasive (Gent, 2012). Abrasive particle size does not significantly affect cutting speed but has the impact on surface roughness (Vajdová, 2014).

Abrasive mass flow—*Ma* [g min⁻¹]

The mass flow of abrasive expresses the amount of the abrasive added per specified time unit. It significantly affects the cutting process. When increasing the mass flow of the abrasive, we can increase the cutting speed, while retaining the quality of the workpiece surface. The amount of abrasive used in practice ranges between 200 and 700 g/min. One of the possible ways to achieve a greater cutting thickness is to increase the number of jet transitions along the same cutting trajectory, without changing other technological parameters (Gent, 2012).

1.4 Quality of Machined Surfaces of Components After AWJC

The essential evaluation criteria of the components are (Bátora & Vasilko, 2000; Kleinedlerová, 2011):

- Length dimensions,
- Angular dimensions,
- Surface roughness,
- Surface defects,
- Quality of machined surface,
- Defects on edges and cut surfaces.

The most frequently evaluated qualitative indicator of the machined surface (on the cut surfaces) is surface roughness (Hloch, 2008). Surface roughness after AWJC is standardly categorized into five areas (STN ISO 4287–2 (014,450), 1993; Kleinedlerová & Janáč, 2013). The research results confirmed that the AWJ has capable also for the production of micro inclined holes than other processing techniques based on lesser machining time and satisfactory hole quality. Micro inclined holes are widely used as cooling holes in the jet engine and gas turbine components for its protection from a high temperature gas stream. The effect of piercing parameters on machining time, hole entry diameter, and surface morphology were examined. From the experimental results, it is confirmed that the inclined hole surfaces do not contain any cracks and delamination. However, the rough morphology, i.e. wear tracks and microchips was seen in the hole wall surfaces (Yuvaraj, 2020).

References

Bátora, B., & Vasilko, K. (2000). Obrobené povrchy (Machined surfaces) (p. 183). Trenčín University.

Folkes, J. (2009). Waterjet-An innovative tool for manufacturing. *Journal of materials processing technology, 209*(20), 6181–6189.

Gent, M., et al. (2012). Experimental evaluation of the physical properties required of abrasives for optimizing waterjet cutting of ductile materials. *Wear, 25*, 43–51.

Hashish, M. (1988). Visualization of the abrasive-waterjet cutting process. *Experimental Mechanics, 28*, 159–169. https://doi.org/10.1007/BF02317567

Hashish, M. (1991). Optimisation factors in abrasive waterjet machining. *ASME journal of Engineering of industry, 113*(1), 29–37.

Hassan, A., & Kosmol, J. (2001). Finite element modeling of Abrasive Water Jet Machining (AWJM). In *Conference: 15th international conference on jetting technology, Ronnby, Sweden* (pp. 321–333). BHR Group 2000 Jetting Technology.

Hlaváč, L., & Vašek, J. (1994). Physical model of high energy liquid jet for cutting rock. *International Journal of Water Jet Technology, 2*(1), 39–50.

Hlaváč, L., Sosnovec, L., & Martinec, P. (1999). Abrasives for high energy water jet investigation of properties. In M. Hashish (ed.), *Proceedings of the 10th American waterjet conference* (pp. 409–418). WJTA.

Hlaváč, L. (2000). *Model pro řízení parametrů kapalinového paprsku při porušování materiálů v pevné fázi.* Doktorská disertačná práce (p. 101). VŠB Technická univerzita.

Hloch, S., & Valíček, J. (2008). *Vplyv faktorov na topografiu povrchov vytvorených hydroabrazívnym delením* (p. 125). FVT TU. ISBN: 978-80-553-0091-7

Hloch, S., et al. (2020). Effect of pressure of pulsating water jet moving along stair trajectory on erosion depth, surface morphology and microhardness. *Wear, 452–453* Article 203278.

Hloch, S. (2007). Influence of nozzle wear on surface quality at abrasive waterjet cutting. In *Annals of DAAAM for 2007 & proceedings of the 18th international DAAAM symposium—intelligent manufacturing & automation* (pp. 329–330).

Kleinedlerová, I., & Janáč, A. (2013). Surface roughness when abrasive water jet cutting. *Strojírenská technologie, 3.* ISBN 978–80–261–0136–9.

Rezmat (2018). *Interné materiály (Internal materials).* REZMAT s.r.o. Dubnica nad Váhom.

Kleinedlerová, I., & Janáč, A. (2012). Meranie deformácie výstupného otvoru abrazívnej dýzy (Deformation measurement of the abrasive nozzle) *Trilobit, 2.* ISSN 1804-1795.

Kleinedlerová, I., Janáč, A., & Kleinedler, P. (2011). Analysis evaluation parameters of Ra, Rz surface roughness on short sectors. *Manufacturing Engineering, 10*(2), 10–13. ISSN 1335-7972.

Krajný, Z. (1998). *Vodný lúč v praxi (Water Jet in practice)* (p. 384). Epos. ISBN 80-8057-091-4

Kulekci, M. K. (2002). Processes and apparatus developments in industrial waterjet applications. *International journal of machine tools and manufacture, 42*(12), 1297–1306. ISSN: 0890-6955.

Maňková, I. (2000). *Progresívne technológie (Progressive technologies)* (p. 275). Vienala. ISBN 80-7099-430-4.

Mardi, B., et al. (2021). AWJM of Mg-based composites with different nanoparticle contents. *International Journal of Advance Manufacturing Technology, 12*.

Nag, A., et al. (2021). Effect of acoustic chamber length on disintegration of ductile material with pulsating water jet. In *Advances in manufacturing engineering and materials II: Proceedings of the ICMEM 2020, Nový Smokovec* (pp. 120–131). Springer International Publishing.

Natarajan, Y. (2020). Abrasive water jet machining process: A state of art of review. *Journal of Manufacturing Processes, 49*, 271–322.

Norma (Standard) DIN 8580: 2003–09 (2003). *Manufacturing processes: Terms and definitions, division* (p. 13).

Ohlsson, L., et al. (1992). Optimisation of the piercing or drilling mechanism of abrasive water jets. In A. Lichtarowicz (Ed.), *Jet cutting technology. Fluid mechanics and its applications* (vol 13). Springer. https://doi.org/10.1007/978-94-011-2678-6_24

Summers, D. A., & Blaine, J. G. (1997). A fundamental tests for parameter evaluation. In *Geomechanics 93* (pp. 321–325). Balkema.

Vajdová, A. (2014). Design and verification of non-conventional technologies carving metal components cars. *Strojírenská technologie, 2*. ISSN 1211–4162.

Vasilko, K., et al. (1990). *Nové materiály a technológie ich spracovania (New materials and technologies of their processing)* (p. 355). Alfa. ISBN 80-05-00661-6

Yuvaraj, N., et al. (2020). Abrasive water jet piercing of inclined holes on ceramic coated nickel superalloy: A preliminary study. *Manufacturing Letters, 26*, 59–63.

Zain, A. M., Haron, H., & Sharif, S. (2011). Optimization of process parameters in the abrasive waterjet machining using integrated SA–GA. *Applied Soft Computing—Journal, 11*(8), 5350–5359. ISSN: 1568-4946.

Zeng, J., & Kim, T. J. (1992). Development of an Abrasive water jet kerf cutting model. In *Proceedings of the 11th International conference of Jet cutting technology* (pp. 483–501). Scotland.

Chapter 2
Piercing by AWJC

Piercing by abrasive water jet is defined as gradual penetration of a jet into the mass of material being machined, over its entire thickness, with the aim of forming a through hole. A groove made by expanding the formed hole creates the primary precondition for further movement and action of the jet in the material being machined. Piercing can also be defined as one of the important strategies of the tool movement during AWJC. In the machined material, the abrasive water jet can prepare a starting position for its further action. The technology of AWJC enables formation of a through hole—a piercing in the entire thickness of the machined material. The resulting opening represents the space in which the jet stabilizes after forming the hole. The jet is then ready to cut the material (Fredin, 2011; Liu, 2006). Piercing of materials is preceded by the initial jet impact which several times exceeds the force acting on the material, compared to the force during the actual cutting. Depending on thickness, the material being machined is pierced at the point of impact, either immediately or in a few seconds. The time interval of the jet action upon its penetration into the workpiece is defined as the exposure time. Measured from the moment the jet impact on the material surface up to its initial exit from the machined material (Fredin, 2011), it expresses the time interval required to form a continuous through hole in the material being machined. Piercing starts as soon as a jet penetrates material. An accompanying feature of piercing is spattered of the mixture from the used jet (its individual components: liquid, abrasive, and air), together with the material particles having been removed, into the vicinity of the piercing point, i.e. toward the sides and against the direction of the penetrating jet action. Both, the used jet and the particles of removed material, are disposed from the workplace by the pressure of the incoming "new" jet. Spraying is the only possible way of removing the above-mentioned working mixture (Liu, 2009). Evidence of piercing, i.e. formation of a continuous hole, is the end of chaotic spattering of the working mixture into the environment, which is due to the formation of a continuous through hole, and subsequently the changed acoustic sound accompanying the cutting process (Akkurt, 2009). Currently, the aim of piercing is to remove the maximum volume of material, given its thickness,

© The Author(s), under exclusive license to Springer Nature Switzerland AG 2022
I. Kleinedlerová and P. Kleinedler, *Piercing of Materials with Abrasive Water Jet*,
SpringerBriefs in Applied Sciences and Technology,
https://doi.org/10.1007/978-3-030-92130-9_2

on the minimum area and in the shortest possible time interval (Fredin, 2011; IGEMS, 2020). To achieve this aim, it is always necessary to consider specific features of the material to be cut.

2.1 Undesirable Phenomena of Material Piercing

Undesirable phenomena of material piercing are usually the following (Matec, 1994– 2003):

Risk of damaging the surface layer of the machined material—the dispersed liquid particles, especially the abrasives directed to the area around the cut, dispose sufficient energy to affect the surface layer of the machined material at the place they fall (Fig. 2.1). The mixture of water droplets and abrasive forming the envelope of the water jet expands with the increasing jet length. The mixture significantly affects the cut material at the place of piercing and its immediate vicinity (Rezmat, 2018), (Matec, 1994–2003).

Prolonged time interval of machining—time interval of the jet penetration into the machined material is expressed as the exposure time. In case of a multiple use (e.g. a higher number of holes), piercing represents a significant factor of increasing the time required to machine the material (Rezmat, 2018; Matec, 1994–2003).

Spatter of individual working mixture components into the environment— brings about an increased wear rate of the equipment parts located in the immediate vicinity (e.g. the cutting head). Simultaneously, individual small particles of the working mixture get to the places of the stationary and moving parts of device, which requires their more frequent inspection (Rezmat, 2018).

Deteriorated quality of the cut surface—when cutting mainly the white and yellow plastics (Erthalon, Silon, Macralon), their original color disappears on the cut surface, which is due to the longer contact with water in the collection container. The original shade darkens and "fades" owing to the contact with used "dirty" water in the

Fig. 2.1 Blasting of the material surface layer during piercing

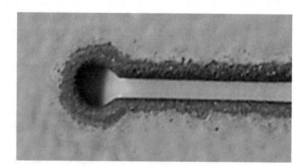

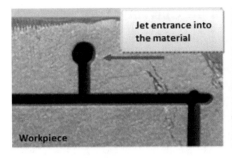

Fig. 2.2 Character of the entrance and exit areas of the jet. *Source* Rezmat, 2018

collection container. Neither rinsing in water nor brushing the material restore the original color (Rezmat, 2018; Matec, 1994–2003).

Increased noise—the cutting noise reaches the average values of 90 dB, and increases up to 115–125 dB during piercing (Rezmat, 2018).

Damaged material around the piercing site—the nature of damage to the material varies. Damage occurs on both sides of the workpiece. In some cases, the material remains undamaged at the side of the jet entrance and is only damaged in the region of the jet exit from the material on the opposite side of the workpiece (Fig. 2.2).

The spatter of individual working mixture components into the environment together with the mixture of water drops and abrasive, which form the envelope of the water jet, results in, e.g. roughening, opacity, loss of the original gloss, blasting, or even removal of the material surface layer around the piercing site (Rezmat, 2018; Matec, 1994–2003).

Decreased efficiency of the jet at the point of piercing—the incoming and outgoing jets interfere in the area of a non-continuous hole. There is an interaction of mutually opposing forces (Rezmat, 2018; Matec, 1994–2003).

Chemical reaction—prolonging the interval of the water effect on the material brings about the chemical reaction of rays and tannins in the material (e.g. in case of cutting wood), which leads to adverse coloration (Rezmat, 2018).

Solution to eliminating the phenomena accompanying the piercing process—In the current practice, certain short comings are addressed, inter alia, by locating the start of cutting outside the workpiece cutting blank.

The problems related to the spatter of the working mixture into the environment are dealt with by using a protective cap on the cutting head (Rezmat, 2018).

2.2   Analysis and Classification of the Currently Used Piercing Methods

OMAX Corporation, which is an AWJ machine manufacturing company, demonstrated four different types of piercing methods including stationary piercing, dynamic piercing, wiggle piercing, and low-pressure piercing. Stationary piercing is the simplest form of piercing method. The nozzle stays in one location while it pierces the part. During dynamic piercing, the nozzle moves toward a cutline while it pierces the material. The wiggle piercing method uses a slight back and forth movement of the nozzle along the cutline while it pierces the substrate and moves toward the cutline after the piercing process. During the low-pressure piercing method, the AWJ pierces and cuts the material with low-pressure water and abrasive (Kim, 2020).

During the experiments giving a lot of information on the effect of piercing parameters on piercing time, it can be concluded that it is of great importance how piercing variables are chosen if a short piercing time is to be achieved. Dynamic piercing is generally faster than stationary piercing. Stationary piercing is only useful when cutting really small holes. Linear piercing is the most efficient piercing method of the ones investigated. For linear piercing as high feed rates as possible, for the given work piece thickness shall be chosen. For thicker work pieces this is a challenge since a too high feed rate will not pierce the material at all but a too low feed rate will increase the piercing times exponentially. Repeated linear piercing and circular piercing is not assensitive against changes in piercing parameters but they are however generally slower. They have the advantage of being useful when smaller holes shall be cut. Experiments on standoff distances show that a slightly larger standoff distance can preferably be chosen for piercing in comparison with cutting, with no risk of increasing the piercing time but with less risk of clogging the nozzle (Fredin, 2011).

In the research (Ohlsson, 1992) two different movement types were compared to stationary jet piercing, these were linear and circular. Optimizing the linear movement piercing is a matter of trying a number of speeds but circular movement is rather more complex. In the case of circular movement during piercing there are two major parameters: the diameter of the circle (D) and the velocity of the movement (v). The diameter of the circle in this case means the CNC programmed diameter of the movement of the jet. The area of influence of the jet will so be a combination of this circular movement and the diameter of the jet itself. Throughout this experiment the diameter of the jet and all other jet related variables were kept constant. Our investigation of circular movement drilling involved changing values for the diameter of the circular movement and the velocity of movement. It was discovered that an optimum range of D and v exists within which it is possible to achieve very short penetration times (Ohlsson, 1992).

Based on the studied sources regarding the topic of the current monograph, the methods of piercing materials by AWJC are classified as follows (Matec, 1994–2003), (Fredin, 2011):

(a) **Continuous piercing** is characterized by a permanent, continuous action of the jet on the material being machined. The jet is characterized by a constant energy level during cutting.

(b) **Discontinuous piercing**: the jet is characterized by an unstable energy level during cutting. Originally a continuous jet is divided into mutually separated columns of water, forming the so-called pulsed jet (Bortolussi, 2009).

The experimental water jet cutting described in this monograph used continuous piercing; the continuous piercing by a continuous abrasive water jet is therefore described in more detail.

Continuous piercing—the classification criteria of continuous piercing are described in (Matec, 1994–2003) and (Fredin, 2011) as follows:

(1) v_f **[mm min^{-1}]**—feed rate of jet on the machined surface of the cut material during piercing:

Stationary piercing: $v_f = 0$ mm min^{-1}—piercing performed by a stationary jet.

Non-stationary piercing: $v_f \neq 0$ mm min^{-1}—a hole performed by a moving jet.

(2) Lp **[mm]**—length of the jet trajectory performed during piercing:

Point-type: $Lp = 0$ mm—the jet acts at one point of the pierced material that does not change over time.

Curve-type: $Lp \neq 0$ mm—the jet acts on a group of points, forming a curve of a moving jet when piercing.

(3) **Overlapping of the criteria 1 and 2 give** (Fredin, 2011):

Stationary point piercing which is performed by a stationary jet acting during the piercing on a non-changing point of the pierced material. The jet performs piercing at a single point located outside the contour shape of the workpiece. After formation of a continuous hole, the jet is set in motion while approaching the contour of the cut workpiece. The size of the surface on which the jet acts, S [mm^2], is identical with the jet diameter. With the jet diameter $d = 1.1$ mm, S is 3.8 mm^2 (Matec, 1994–2003).

Non-stationary curve piercing is performed by a moving jet acting on the group of points forming the trajectory of the moving jet curve during piercing. The jet strikes the surface of the material and moves at a speed of v_f toward the contour shape of the workpiece. The process of material removal takes place during the movement of the jet toward the cut part (Matec, 1994–2003).

According to the shape of a moving jet trajectory, we distinguish (Matec, 1994–2003; Fredin, 2011).

Non-stationary linear piercing takes place when the jet trajectory is a straight line. The size of the area on which the jet acts, S [mm^2], is determined by the product of the trajectory length and the jet diameter. The changing parameter is the v_f one (Matec, 1994–2003). **Non-stationary non-linear piercing** takes place if the jet trajectory

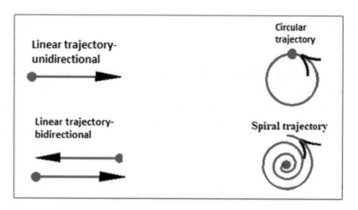

Fig. 2.3 Trajectory of a moving jet during piercing. *Source* Rezmat, 2018

takes a shape other than a line (arc, circle, spiral). The size of the surface on which the jet acts, S [mm^2], is determined by the diameter of the circular trajectory along which the jet moves. The changing parameter is the diameter of the circular trajectory (Matec, 1994–2003). The shape of the moving jet trajectory during piercing can thus vary.

Figure 2.3 shows certain possible ways of the jet movement during piercing (Matec, 1994–2003).

2.3   Possibilities of Piercing Materials by AWJC, Using Current Software

The current best-known suppliers of machine tools and software for AWJC are, e.g.:

- IGEMS, Sweden,
- **FLOW International Corporation, USA,**
- **OMAX Corporation, USA,**
- WATER JET SWEDEN AB, Sweden,
- WARDJet, USA,
- MAXIEM Waterjets, England,
- WRYKRYS, Czech Republic.

They offer the following options of piercing the materials using the AWJC technology (IGEMS, 2020):

- **Blind lead**—provides a special option of piercing blind holes,
- **Linear piercing**—refers to non-stationary linear piercing,
- **Stationary piercing**—refers to stationary point piercing,
- **Circular piercing**—refers to non-stationary curve (circular) piercing,
- **Drilling**—refers to piercing by drilling, (used for laminate materials in particular),

- **Air start**—is used to make a piercing outside the material, instead of in the semi-product in order to cut workpieces. It can be applied only for particular workpieces of the shape enabling it,
- **User piercing**—used for piercing according to the user specific setting requirements.

In order to perform experimental cuts within the research described in the current monograph, the REZMAT s.r.o. Company in Dubnica n/V, where the experiments were conducted, used the available **WRYKRYS** software. The WRYKRYS software provides an option of designing the following ways of piercing (WRYKRYS):

- Stationary point piercing,
- Non-stationary linear piercing,
- Non-stationary non-linear (circular) piercing.

The above-mentioned methods of piercing holes in materials enable to modify the basic dimensions of the pierced openings (such as the angle and length of the piercing trajectory) upon request.

A survey into the current available software (listed above) on the domestic and foreign markets showed that piercing the material with the aim of making a through hole, as a pre-requisite for leading the jet to the contour of the cut workpiece, can be performed only using the software and types of perforations mentioned in this monograph, always in the space of the residual material.

2.4 Piercing in Materials with ACJ and FAWJ

The first FAWJ (Fine Abrasive Water Jet; Micro waterjet; Finecut) cutting head was developed by Water Jet Sweden in 2008. The FAWJ cutting process requires very fine abrasives of 220–245 mesh and a precise CNC controlled abrasive feeder (Waterjet Sweden, 2021). Piercing damage to target materials results in the form of surface/subsurface cracking, chipping, and delamination as long as their ultimate strengths are weaker than the piercing pressure buildup inside blind holes. As the jet approaches the bottom of blind holes, the fluid decelerates, stops, and reverses it course and then exits through the hole entrance. During that process, kinetic energy of the jet converts into potential energy and the piercing pressure develops inside the blind hole. The magnitude of the piercing pressure is inversely proportional to the local speed of the water. By definition, the maximum piercing pressure is the stagnation pressure at which the speed of the water vanishes (Liu, 2009). ACJ (Abrasive cryogenic jet) and the FAWJ demonstrat their advantage over the AWJ for mitigating piercing damage, the FAWJ is considerably less bulky, more cost-effective and user-friendly, and safer to operate than the ACJ. The FAWJ is expected to be the missing link to help fully realize the potential of waterjet technology as a truly material-independent, precision machine tool. Such a unique property would lead to the development of a one-inall and all-in-one machine tool for all materials. This would further promote the growth of waterjet technology for machining advanced materials. A prototype FAWJ is currently under fabrication and testing. FAWJ stands for, emulates the performance of the ACJ for mitigate piercing damage through rapid evaporation of the working fluid upon exiting the orifice and mixing tube. Only a small fraction of the water in the FAWJ enters the blind hole, resulting in a significant reduction of the piercing pressure. The structural integrity of the workpiece is preserved so long as its ultimate strength is higher than the piercing pressure (Liu, 2009).

In research aimed on improving piercing quality through three independent but related experiments regarding piercing with an AWJ and a FAWJ in borosilicate glass were reached listed conclusions (Schwartzentruber, 2015):

– Higher pressures produced significantly larger defects. Use of a lower pressure reduced the probability of fracturing upon impact and induced a gentler erosion action.
– Increasing the SOD (standoff distance) had a significant effect on hole size and hole aspect ratio with a trend indicating that increasing SOD increased size while improving aspect ratio.

The second experiment used stacked glass plates during FAWJM piercing operations in order to reduce the amount of exit chipping. The stacking experiment was simulated as a FE analysis in ANSYS. Comparison between FE analyses of the stacking configurations showed a significant stress reduction at the point of jet exit which, in turn, reduced the amount of chipping on the exit pierce. Lastly, a novel FAWJ/AJM hybrid process was presented and found to greatly decrease both exit hole size and damage, albeit at the expense of machining time (Schwartzentruber, 2015).

The chipping was reduced by using the less aggressive erosion mechanisms of AJM to finish FAWJ machined blind holes. Due to the versatility of both AJM and AWJ, this process can be used to machine a wide range of brittle and hard to machine materials that would be difficult otherwise.

The results reported have significantly reduced the amount of entrance and exit chipping AWJM produces in borosilicate glass. The practices presented in this paper may also be applicable to other brittle materials, and thus expanding the versatility and industrial application of AWJM/FAWJM (Schwartzentruber, 2015).

2.5   Piercing in Composite Materials

Delamination is a major concern in the manufacturing processes of composite materials. It reduces not only the structural integrity of the laminate but also the long-term reliability of the assembly. Water jet drilling, in spite of its advantages of no tool wear and thermal damage, often creates delamination composite laminate at bottom. The current paper presents an analytical approach to study the delamination during drilling by water jet piercing. The analysis uses fracture mechanics with plate theory to describe the mechanism of delamination. This model predicts an optimal water jet pressure for no delamination as a function of hole depth and material parameters (opening-mode delamination fracture toughness and modulus of elasticity). Good agreement is achieved with data obtained from water jet drilling of graphite epoxy laminate. The predicted optimal water jet pressure can be applied in a control scheme for maximizing the productivity of water jet drilling of composite laminates (Ho-Cheng, 1990).

Montesano et al. compared the drilling quality between a composite laminate test specimen machined with traditional machining and AWJ machining. Both reported that they were able to detect more damage on the test specimen machined with traditional machining (Kim, 2020; Montesano et al., 2017).

By definition, composite laminates contain multiple layers, which makes interlaminar bonding strength a concern during localized impact of an AWJ stream. Delamination between layers is one of the most common defects caused by AWJ machining of composite material. The occurrence of delamination on composite laminates during the waterjet machining process can be described in two steps; initial crack generation and crack propagation. For the initial crack creation, it was explained that the shock wave generated by the waterjet stream impacting the surface of the workpiece causes the initial crack. It was described that the initial crack is caused by a water wedge effect as the waterjet penetrates the workpiece. Then both agreed that the water and abrasive mix penetrates the crack tip and propagates the crack due to hydrodynamic pressurization during the cutting process. Since delamination of composite laminates degrades mechanical properties of the structure, it needs to be considered during machining process planning in order to be minimized. Many researchers have demonstrated that delamination of composite laminates during AWJ machining process is influenced by AWJ machining parameters; water pressure, standoff distance, delay and event timing, mixing tube diameter, focusing nozzle diameter, traverse speed, and abrasive mass flow rate. The delamination during the AWJ machining process cannot be fully avoided but reduced with optimizing the machining parameters and conditions. The Research showed that the water pressure and standoff distance affect delamination of composite materials more than the abrasive mass flow rate during the AWJ machining process. Speed during machining of AWJ affects the size of delamination the most. The delamination can be reduced by reducing the water pressure and the size of the jet diameter (Kim, 2020).

References

Akkurt, A. (2009). The effect of material type and plate thickness on drilling time of abrasive water jet drilling process. *Materials and Design, 30*, 810–815.

Bortolussi, A. et al. (2009). Ornamental stone surface treatment by pulsating water jets. In *Proceedings of 9th Pacific Rim International Conference on Water Jetting technology, Koriyama-city, Japan* (pp. 189–193).

Fredin, J., & Jönsson, A. (2011). Experimentation on piercing with abrasive water jet. In *World academy of science, engineering and technology* (Vol. 59, pp. 2611–2617).

Ho-Cheng, H. (1990). A failure analysis of water jet drilling in composite laminates. *International Journal of Machine Tools and Manufacture, 30*(Issue 3), 423–429.

IGEMS, Sweden. (2020). *CAD/CAM/NEST. Reference manual* (p. 207).

Kim, G., Denos, R. B., & Sterkenburg, R. (2020). Influence of different piercing methods of abrasive waterjet on delamination of fiber reinforced composite laminate. *Composite Structures, 240.*

Liu, H. T. (2006). Empirical modeling of hole piercing with abrasive waterjets. *International Journal of Advanced Manufacturing Technology, 42*, 263–279.

Liu, H. T., & Schubert, E. (2009). Piercing in delicate materials with abrasive-waterjets. *International Journal of Advance Manufacturing Technology, 42*, 263–279.

Matec. (1994–2003). *Interné materiály (Internal materials) ZTS – MATEC a.s. Dubnica nad Váhom* (p. 114).

Montesano, J., et al. (2017). Influence of drilling and abrasive water jet induced damage on the performance of carbon fabric/epoxy plates with holes. *Composite Structures, 163*, 257–266.

Ohlsson L., et al. (1992). Optimisation of the piercing or drilling mechanism of abrasive water jets. In: A. Lichtarowicz (Eds.), *Jet cutting technology. Fluid mechanics and its applications* (Vol. 13). Springer. https://doi.org/10.1007/978-94-011-2678-6_24.

Rezmat. (2018). *Interné materiály (Internal materials) REZMAT s.r.o. Dubnica nad Váhom.*

Schwartzentruber, J., & Papini, M. (2015). Abrasive waterjet micro-piercing of borosilicate glass. *Journal of Materials Processing Technology, 219*, 143–154.

Waterjet Sweden. (2021). Fine abrasive waterjet compliments existing technologies © copyright, 2021, wjs uk ltd [cit. 2021–05–20]. https://www.waterjetsweden.co.uk/water-jet-cutting-news/micro-abrasive-waterjet.

Chapter 3
Design of an Innovative Approach to Piercing the Material

To attain production of a desired shape, the actual working movement of the jet along the determined contour of a workpiece is preceded by the approach of the jet to the contour.

The key factors determining the jet's approach strategy to the workpiece contour are (according to IGEMS, 2020), e.g.:

- Type of the material to be machined,
- Condition of the material to be machined (e.g. flatness, former method of processing the semi-finished product),
- Thickness of the material,
- Condition of the material surface to be machined (e.g. in case of polished surface, it is necessary to locate the piercing point sufficiently far from the cut contour, owing to the local sandblasting into the surroundings of the pierced point owing to the action of abrasive during the period of piercing, etc.),
- Size of the area on which piercing would be performed,
- Required resulting quality of the machined surface,
- Contour shape of the cut workpiece,
- Considered future shape of the jet trajectory.

Implementation in Practice:

Designers of an NC program carefully inspect the material blank to be cut. They get acquainted with the type and thickness of the material as well as state of the material surface quality. They precisely measure its dimensions or draw its shape. The measured dimensions and shape are then entered into a graphics program. Subsequently, the designers carry out positioning of the workpiece in the drawn shape of the semi-finished product with the aim of optimizing the spatial utilization of the material to be cut, and determining the options of positioning and manipulating the cut workpieces of the supplied semi-finished material. After positioning the workpieces, they consider the approaches to the contours of the cut workpieces. The

I. Kleinedlerová and P. Kleinedler, *Piercing of Materials with Abrasive Water Jet*, SpringerBriefs in Applied Sciences and Technology, https://doi.org/10.1007/978-3-030-92130-9_3

Fig. 3.1 Completed piercing
in the residual material

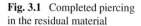

above-mentioned factors represent the starting point of designing an NC program
containing the future method of manufacturing, the shape of the approach trajectory and the definition of a part of the technological parameters in a digital form.
Software helps programmers to design approaches. The software programs supplied
with the technology provide the users with specific, comprehensive, and predefined
procedures regarding the piercing methods with the aim of minimizing the operator's
requirements. After entering the input data, a particular program designs the complete
procedure of piercing, including the assignment of technological parameters. After
starting the device, the designed NC program is applied to the particular material, in
order to provide conditions for approach to the contour of the workpiece to be cut.
After piercing the material and the subsequent formation of a hole, the jet gradually transits from the point of piercing to the contour shape of the workpiece being
cut. The transition rate is determined primarily by the material type and thickness.
In general, the transition rate significantly decreases with the increasing material
thickness. As an example, I can present the initial state of piercing a DIN 1.4301
stainless steel material of a 30 mm thickness. After the initial impact on the material
($v_f = 0$), the jet moves along the software-designed technological parameters, i.e.
the track of 15 mm and the speed $v_f = 10$ mm min^{-1}. After piercing the material,
the jet continues its movement at an unchanged velocity toward the contour of the
workpiece to be cut. After approximating a distance of about 2 mm from the workpiece contour, it decelerates to a velocity of $v_f = 5$ mm min^{-1}. Such a velocity is also
used for transition to the contour of the workpiece and the subsequent trajectory of
about 3 mm. Then, the speed gradually increases up to the so-called working speed,
conditioned by the specified required quality of the machined surface and the workpiece being cut. Characteristics of the piercing described above: the piercing was
located in the residual material (Fig. 3.1), the shape of the piercing trajectory was of
the non-stationary linear piercing type. The course of piercing was accompanied by
a double divergence of the jet. In practice, the entrance into the material is performed
mainly on the basis of the possibilities provided by software.

According to the literature survey and common practice, the initial state of the issue can be characterized as follows:

The approach to the material should generally ensure:

- Formation of a through hole in the entire volume of the machined material at the shortest possible distance and in the shortest time,
- Avoiding damage to the machined material at the point of transition, approach to the contour of the cut workpiece from the space in which the piercing was performed.

The software applications supplied with the technology offer a wide range of options of the jet approach to the contour of the cut workpiece. Despite variety of all methods designed by software and used in practice to date, there is always one common element: performance in the space of the residual material. That means that the initial entry of the jet into the material is located in the space outside the workpiece to be cut. From this space, various variants of geometric shapes such as lines, curves, tangential and radius connections, etc., are conducted toward the workpiece with the aim of connecting the piercing points in the residual material with the point of piercing transition to the contour of the workpiece being cut. Since the current technology enables, to a certain extent, cutting all types of materials, frequently very expensive and relatively rare ones, an increased attention is paid in practice to saving the volume of residual material.

The following part of the monograph presents the author's original design of piercing implementation. Piercing of material with the aim of making a continuous hole as a pre-requisite for the jet approach to the contour of the workpiece being cut is currently always performed in the space of the residual material. As part of my own innovative design, I experimentally investigated the option of piercing a material directly on the workpiece contour.

The research was supplemented by an original mathematical formula determining the basic piercing parameters.

Reference

IGEMS, Sweden. (2020). *CAD/CAM/NEST. Reference manual* (p. 207).

Chapter 4
Experimental Research

Experimental research into cutting was conducted in the REZMAT s.r.o. Company in Dubnica nad Váhom. The Company performs cutting by both, a clean and abrasive water jets, while also supporting research in the area.

4.1 Experimental Conditions

Experimental cutting was performed on a *WJxxyy-nZ-D* machine (Table type D) by the PTV Company. It is a machine of a relatively simple design, with a table size ranging from 1×1 m to 4.5×12 m. The machine can be used for cutting by a clean as well as abrasive water jet, with the option of using two cutting heads and the maximum possible Z-axis stroke up to 300 mm.

To perform experimental cuts in the REZMAT s.r.o. Company in Dubnica n/V, we used the available WRYKRYS software provided by the Company. The software allows to write a production NC program which can then be modified according to particular demands. It was essential to enable additional insertion of a subprogram for making a piercing on the workpiece contour into the generated text file of the NC program, and thus program the required jet trajectory during piercing. Within the experiments, a commonly available DIN 1.4301 material was cut in the range of the selected thicknesses of 15–80 mm.

4.1.1 Technological Parameters of Cutting for Piercing on the Workpiece Contour

In Phase 1 of attaining the goals, it was necessary to select, within the experimental conditions, technological parameters of AWJC, in order to make a hole, as well as to

I. Kleinedlerová and P. Kleinedler, *Piercing of Materials with Abrasive Water Jet*, SpringerBriefs in Applied Sciences and Technology, https://doi.org/10.1007/978-3-030-92130-9_4

prepare a set evaluation criteria to continually assess and evaluate the attained results. During the study and subsequent experimental work with technological parameters directly in practice, considered were the effects of technological parameters of cutting. Some of the parameters provided virtually no possibilities of modification, e.g. the maximum fluid pressure p (limited by the AWJC equipment used), while others allowed yet greater extent of alteration, e.g. when working with an abrasive, or in case of the focusing nozzle mouth distance from the machined material.

The greatest attention was paid mainly to those parameters that could be directly determined and adjusted based on the requirements such as the machined material, demands of practice, conditions in individual workplaces, and their availability.

Selected technological parameters of cutting:

Liquid pressure: When choosing the liquid pressure, the theoretical recommendations supported by general practice were taken into account, e.g. the higher the pressure, the better the cutting results. In the experimental part we therefore used the maximum working pressure. In the given operation, the manometer displayed the working pressure in the range of 600 MPa.

The range of liquid pressure is given by its partial pulsation after compression, which arises as a result of the action of a double-acting multiplier in its boundary positions during the liquid compression.

Abrasive material: Garnet abrasive is used almost exclusively in practice for AWJC. The experiments were therefore performed with the most commonly used abrasive Garnet, MESH 80 (grain size within 180–220 μm).

Abrasive mass flow: Based on the results obtained and the above-mentioned facts, the amount of 600 g.min^{-1} was chosen.

Distance of the focusing nozzle mouth from the cut material: The distance of the nozzle mouth from the material being machined must be as small as possible in order to avoid unnecessary propagation of the jet and the associated dissipation of the jet energy into the surroundings. Regarding the above-mentioned, the selected distance of the focusing nozzle mouth from the material to be machined for further experiments was 2 mm.

4.1.2 Shape of the Jet Trajectory During Piercing the Material

In addition to the standard technological parameters, shape of the jet trajectory during the piercing itself was also taken into account. In order to make a piercing on the workpiece contour, targets have been set according to the observed practices, resulting from the required properties, which should take into account the trajectory shape used:

1. Ensure constant contact of the jet with the material.

2. Provide smooth connection to the previous jet trajectory.
3. Provide a smooth jet approach to and exit from cutting when passing the contour geometry.

- Objective of points 1–3: to eliminate possible manifestations of the jet action in the material space outside the cutting gap.

4. To eliminate the jet movement into the completely empty space within the cutting gap, while taking into account the extent of the mutual degree of overlap of the jet trajectory.

- Objective: Time-saving when cutting.

5. Eliminate a sudden halt or stand-still of the jet movement. The consequence of this phenomenon is the jet skipping and uncut area. The consequences worsen with increased thickness of the material being machined.
6. Eliminate a sudden change in the shapes of trajectory orientation to another direction.
7. Use principally rounded radius transitions.

- Objective of points 5–7: to reduce the risk of uncut areas resulting from a phenomenon known as jet skipping.

8. The jet movement along the trajectory should ensure the removal of the entire required material volume from the cutting gap.
9. Respecting the initial jet diameter of 1.1 mm.
10. Smooth movement of the jet along the trajectory without deceleration.

- Objective of point 10: time savings during cutting and programming; deceleration further increases the width of the cutting gap locally.

11. Preserving a certain selected width of the cutting gap, i.e. determined limits.

- Objective of point 11: saving the volume of material being machined

12. Trajectory should be geometrically definable.

- Objective of point 12: flexible modification of the trajectory parameters, possible repeatability of the process.

13. Jet movement must not damage shape of the workpiece being cut.

These recommendations are based on a set of observations and the results obtained from specific and targeted applications, given by, e.g. type and thickness of the machined material and monitored results.

4.1.3   Determination of Piercing Criteria on the Workpiece Contour

When performing piercing directly on the workpiece contour, the latter should meet the following criteria:

- **Making a continual hole in the entire volume of the machined material,**
- **Avoiding damage to the machined surface directly in the place of piercing** through the entire thickness of the material being pierced, beyond the permitted possibilities in connection with other prescribed (considered) operations of machining the given part within a specific technological process beyond the allowed options,
- **Avoiding damage to the machined material surface at the piercing place** beyond the space of cutting gap (i.e. avoiding the damage to the workpiece surface and simultaneously also the surface of residual material).

4.2   Introductory Experiments

For substantial advancement, several initial piercing operations into the DIN 1.4301 stainless steel of the 30 mm material thickness were performed by the AWJC equipment at various settings of technological parameters.

Figure 4.1 illustrates and compares the cut samples with a standard piercing and those with a piercing on the workpiece contour.

The standard material piercing in Fig. 4.1a, c, extends into the residual material (the workpiece shape was damaged).

The piercing on the workpiece contour was performed without damaging the residual material (the workpiece shape was not damaged), as documented in Fig. 4.1b, d.

Figure 4.2 shows the proposed course of piercing the material, including an illustration of the typical manifestation of the jet action on the material being machined by piercing. Such piercing can be characterized as a piercing with non-stationary linear movement of the jet.

Piercing was performed at a distance of L_p, with the jet moving at a velocity of $v_f > 0$, liquid pressure p, and jet diameter of 1.1 mm, with the jet containing a certain amount of the most widely used type of abrasive material. The jet affected a certain type of material of the thickness, during the period, to attain a certain degree of quality of the machined surface.

As also illustrated in Fig. 4.3, the jet moving at a velocity v_f along the machined surface of the material to be cut removes the material at the points of direct contact, subsequently forming a cutting gap in the jet direction, i.e. forwards and downwards.

The jet velocity along the machined surface of the material being cut determines the size of the hole being formed and the angle of the jet incidence on the material.

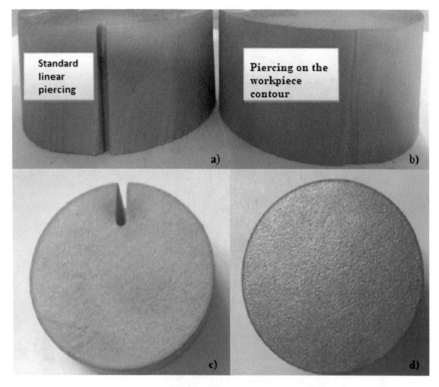

Fig. 4.1 Completed piercing. **a** Standard piercing—side-view, **b** Piercing on the workpiece contour—side-view, **c** Standard piercing—top view, **d** Piercing on the workpiece contour—top view

The moving jet gradually removes an increasing volume of material in the direction of its thickness.

Removal of the material along its entire thickness forms a space of piercing the total volume expressed by the thickness of the material. A continuous hole formed is a basic pre-requisite for further movement of the jet in the given material.

4.3 Mathematical Formula Determining the Piercing Length in the DIN 1.4301 Material

To devise the mathematical formula, it was first necessary to measure the trajectories lengths where a standard linear piercing would be made for the determined material thicknesses and the jet feed rates along the machined surface of the cut material during piercing. The materials considered for cutting were duralumin, stainless steel, and other types of steel (classes 10–15).

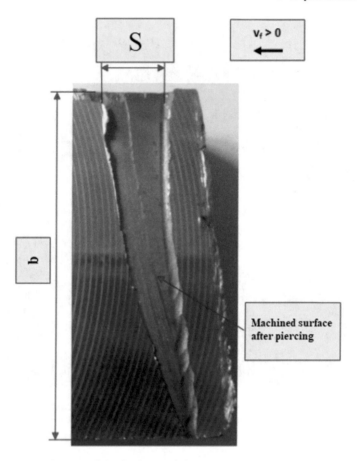

Fig. 4.2 A detail of a place of piercing in a workpiece

Fig. 4.3 Curve of the jet action during non-stationary piercing. *Source* Ohlsson (1992)

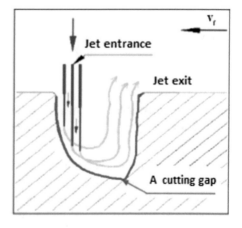

DIN 1.4301 Stainless steel was selected for experimental cutting, since our own Internet survey and several authors (Hloch, 2008) confirmed it was the most common stainless steel material machined by AWJC.

The cutting was performed under the following technological conditions:

Equipment:

WJxxyy-nZ-D (Desk, type D), by PTV Company

High-pressure pump: FLOW 9x-Double

Abrasive system: FLOW PASER III

CNC system: MEFI

Type of the machined material: Stainless steel DIN 1.4301

Thickness of the machined material: 15, 20, 25, 30, 40, 50, 60, 70, 80 mm

Liquid pressure: 400 MPa

Diameter of water jet: 0.33 mm

Diameter of focusing nozzle: 1.1 mm

Abrasive: Garnet, Mesh 80 (average grain size 180–220 μm)

Abrasive mass flow: 600 g min^{-1}.

Length of focusing nozzle: 70 mm

Distance of nozzle from material: 2 mm

Angle of the jet impact on material: 90°

Jet feed rate along the machined surface of the cut material during piercing: 5, 10, 15, 20, 30, 35 [mm·min^{-1}]

Jet feed rate along the machined surface of the cut material—working speed: 160–210 [mm·min^{-1}]

Evaluation criteria:

The length of the resulting material piercing [mm] (Fig. 4.4) for the specified thicknesses of the material to be machined 15, 20, 25, 30, 40, 50, 60, 70, 80 [mm] and the jet feed rate along the machined surface of the material to be cut during piercing 5, 10, 15, 20, 25, 30, 35 [mm·min^{-1}].

The piercing length for each material thickness was evaluated as the arithmetic mean of the five measured length values. The piercing length was measured by a digital calliper, which was sufficient for the given purposes.

The results of the arithmetic means of the measured piercing lengths are recorded in Table 4.1. The values are rounded to one decimal place.

Based on the evaluated piercing lengths, a mathematical formula determining the piercing length for certain material thicknesses of DIN 1.4301 was developed (Table 4.1). The formula was subsequently applied in programming the process of material cutting and the associated modification of the NC production program. Based on the measured piercing lengths of the material (Table 4.1), the formula was developed as follows:

- In the case of all the mentioned thicknesses of materials, the length of pierced material increases with the increasing jet feed rate along the machined surface of the material being pierced. Using mathematical calculations, the resulting shapes of curves were substituted (Tables 4.2 and 4.3) by regression lines formed as linear approximations of the original curves.

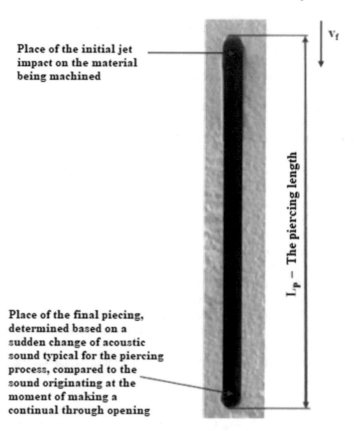

Place of the initial jet impact on the material being machined

Place of the final piecing, determined based on a sudden change of acoustic sound typical for the piercing process, compared to the sound originating at the moment of making a continual through opening

L_p – The piercing length

v_f

Fig. 4.4 Determination of the piercing length

Table. 4.1 Arithmetic mean of the measured values of piercing lengths

	5	10	15	20	25	30	35
Material thickness b [mm]	Arithmetic mean of the measured piercing lengths—L_p [mm]						
15	2.60	3.00	3.20	3.60	3.90	4.10	7.20
20	3.20	3.60	3.90	4.20	4.50	4.80	8.10
25	3.50	4.10	4.30	4.70	4.90	5.10	11.40
30	4.50	4.90	5.40	5.80	6.20	6.40	12.70
40	5.10	5.90	6.30	6.90	7.10	13.90	16.70
50	6.30	7.20	8.20	8.80	9.30	19.30	23.60
60	7.10	8.50	9.40	9.90	10.50	22.70	29.40
70	8.30	10.10	11.30	12.20	25.80	31.30	Uncut
80	9.90	11.50	12.90	14.20	32.70	39.60	Uncut

- That means, the measured piercing lengths were used to calculate the approximate piercing lengths (Table 4.3) according to the line equation: $y = k\,x + q$.

- Based on the slope of approximation lines, we decided to approximate them into one formula expressing the relationship of nine lines (nine machined material

Table 4.2 Calculation of coefficients k and q for the original curves and the regression lines determination

Thickness of material b [mm]	Original curve		Regression line	
	Formula $L_p = k\cdot v_f + q$ v_f—feed rate of jet along the machined surface of the cut material during piercing [mm·min^{-1}]		Formula $L_p = k\cdot v_f + q$ v_f—feed rate of jet along the machined surface of the cut material during piercing [mm·min^{-1}]	
	k	q	k	q
15	0.06057	2.34000	0.05794	2.4515
20	0.06286	2.93333	0.06188	2.932
25	0.06171	3.35333	0.06785	3.4125
30	0.07886	4.15333	0.07585	3.893
40	0.09257	4.84667	0.09794	4.854
50	0.13371	5.89333	0.12815	5.815
60	0.16571	6.60000	0.16648	6.776
70	0.21486	7.70667	0.21293	7.737
80	0.26571	8.76667	0.26750	8.698

Table 4.3 Calculation of approximate length of piercing L_p [mm] to determine regression lines

Thickness of material b [mm]	L_p for $v_f =$ 5	L_p for $v_f =$ 10	L_p for $v_f =$ 15	L_p for $v_f =$ 20	L_p for $v_f =$ 25	L_p for $v_f =$ 30
	v_f—feed rate of jet along the machined surface of the cut material during piercing (mm·min^{-1}) L_p—approximate length of piercing [mm]					
	5	10	15	20	25	30
15	2.64	2.95	3.25	3.55	3.85	4.16
20	3.25	3.56	3.88	4.19	4.50	4.82
25	3.66	3.97	4.28	4.59	4.90	5.20
30	4.55	4.94	5.34	5.73	6.12	6.52
40	5.31	5.77	6.24	6.70	7.16	7.62
50	6.56	7.23	7.90	8.57	9.24	9.90
60	7.43	8.26	9.09	9.91	10.74	11.57
70	8.78	9.86	10.93	12.00	13.08	14.15
80	10.10	11.42	12.75	14.08	15.41	16.74

Table 4.4 Quadratic-linear approximation of piercing length L_p [mm] for one line

Material thickness b [mm]	L_p for $v_f =$ 5	L_p for $v_f =$ 10	L_p for $v_f =$ 15	L_p for $v_f =$ 20	L_p for $v_f =$ 25	L_p for $v_f =$ 30
	v_f—feed rate of jet along the machined surface of the cut material during piercing [mm·min^{-1}] L_p—approximated length of piercing [mm]					
	5	10	15	20	25	30
15	2.74	3.03	3.32	3.61	3.90	4.19
20	3.24	3.55	3.86	4.17	4.48	4.79
25	3.75	4.09	4.43	4.77	5.11	5.45
30	4.27	4.65	5.03	5.41	5.79	6.17
40	5.34	5.83	6.32	6.81	7.30	7.79
50	6.46	7.10	7.74	8.38	9.02	9.66
60	7.61	8.44	9.27	10.11	10.94	11.77
70	8.80	9.87	10.93	12.00	13.06	14.12
80	10.04	11.37	12.71	14.05	15.39	16.72

thicknesses). For that purpose, we used the 2nd degree polynomial substitution which enabled calculation of the final values: a_1s_2, b_1s, c_1, a_2s, and b_2. The "Interpol" mathematical approximation program was used to calculate the values. The final curve of coefficient $k·v_f$ exhibited the shape of a parabola; it was therefore translated by a quadratic approximation. The final curve of coefficient q manifested the linear shape; it was therefore translated by the calculated linear approximation. Subsequently, a quadratic-linear approximation of the piercing length was calculated for the curve corresponding to particular material thickness, as well as the jet feed rate along the machined surface of the material cut by piercing (Table 4.4).

- Based on the calculated values, the procedure for calculating the piercing length was as follows:

Linear approximation of the piercing length L_p [mm] for individual material thicknesses (interval 15 mm to 80 mm), depending on the jet feed rate on the machined surface of the cut material during piercing v_f [mm min^{-1}], according to the measured data:

$L_p = k·v_f + q$ [mm], where
v_f—jet feed rate on the machined surface of the cut material during piercing [mm·min^{-1}].

Quadratic approximation of coefficient k for individual material thicknesses (interval of 15–80 mm) for Formula No. 1:

$k = a_1.b^2 + b_1.b + c_1$, where
$a_1 = 4.06·10^{-5}$

$b_1 = -0.000633$
$c_1 = 0.0583$
$k = 4.06 \cdot 10^{-5}, b^2 — 0.000633, b + 0.0583$

Linear approximation of coefficient q for individual material thicknesses (interval of 15–80 mm) for Formula No.1:

$q = a_2 \cdot b + b_2$, where
$a_2 = 0.0961$
$b_2 = 1.01$
$q = 0.0961 \cdot b + 1.01$

The resulting approximation of the piercing length L_p [mm], determining the minimal piercing length for the DIN 1.4301 material in its thickness range of 15–80 mm:

$L_p = (a_1 \cdot b^2 + b_1 \cdot b + c_1) \cdot v_f + (a_2 \cdot b + b_2)$ [mm],

$L_p = (4.06.10^{-5}.b^2 - 0.000633.b + 0.0583).v_f + (0.0961.b + 1.01)$ **[mm]**, where

v_f—jet feed rate along the machined surface of the cut material during piercing [mm·min^{-1}], 5 mm·min^{-1} to 30 mm·min^{-1},

b—material thickness [mm] ranging from 15 to 80 mm.

The resulting approximation of the jet feed rate along the machined surface of the material cut by piercing v_f [mm·min^{-1}]:

$$v_f = \frac{L_p - (0.096.b + 1.01)}{\left(4.06 \cdot 10^{-5} \cdot b^2 + 0.000633 \cdot b + 0.0583\right)} \left[\text{mm} \cdot \text{min}^{-1}\right]$$

- The Formulas were entered into the MS EXCEL program, which, after entering the thickness of machined material and the jet feed rate along the machined surface of the cut material to be pierced (or after entering thickness of the material to be pierced) allows to calculate the required piercing length (or the of jet speed rate along the machined surface of the to be pierced). Fast calculation of the value and the development of the database of calculations are undisputable advantage.
- Retrospective check of the piercing length calculation according to the revealed deviations in several lengths. The revealed deviations of values were due to a standard approximation error causing the higher of lower calculated approximation values. Such an error can be eliminated, e.g. by performing several measurements of penetration lengths, from, i.e. in the velocities range of 3 mm·s^{-1}–35 mm·s^{-1}, thus obtaining more curve points of approximation. Calculation of the approximated values will be more precise, and calculation of the approximation will thus get closer to the real measured values of the required velocities range of 5 mm·s^{-1}–30 mm·s^{-1}. Table 4.5 illustrates the difference in the measured and calculated values of piercing lengths L_p [mm].
- Identical approach would be applied to calculating an approximation error for the value of v_f (jet feed rate along the machined surface of the cut material [mm·min^{-1}].

Table 4.5 Difference between the measured and calculated approximation values of piercing lengths L_p [mm]

Vf by piercing [mm·min⁻¹]	5			10			15		
Material thickness [mm]	Measured value	Approximate value	Δ measured/approximate value	Measured value	Approximate value	Δ measured/approximate value	Measured value	Approximate value	Δ measured/approximate value
15	2.6	2.74	0.14	3	3.03	0.03	3.2	3.32	0.12
20	3.2	3.24	0.04	3.6	3.55	−0.05	3.9	3.86	−0.04
25	3.5	3.75	0.25	4.1	4.09	−0.01	4.3	4.43	0.13
30	4.5	4.27	−0.23	4.9	4.65	−0.25	5.4	5.03	−0.37
40	5.1	5.34	0.24	5.9	5.83	−0.07	6.3	6.32	0.02
50	6.3	6.46	0.16	7.2	7.1	−0.1	8.2	7.74	−0.46
60	7.1	7.61	0.51	8.5	8.44	−0.06	9.4	9.27	−0.13
70	8.3	8.8	0.5	10.1	9.87	−0.23	11.3	10.93	−0.37
80	9.9	10.04	0.14	11.5	11.37	−0.13	12.9	12.71	−0.19

Vf by piercing [mm·min⁻¹]	20			25			30		
Material thickness [mm]	Measured value	Approximate value	Δ measured/approximate value	Measured value	Approximate value	Δ measured/approximate value	Measured value	Approximate value	Δ measured/approximate va ue
15	3.6	3.61	0.01	3.9	3.9	0	4.1	4.19	0.09
20	4.2	4.17	−0.03	4.5	4.48	−0.02	4.8	4.79	−0.01
25	4.7	4.77	0.07	4.9	5.11	0.21	5.1	5.45	0.35

(continued)

Table 4.5 (continued)

Vf by piercing [mm·min^{-1}]	5			10			15		
Material thickness [mm]	Measured value	Approximate value	Δ measured/approximate value	Measured value	Approximate value	Δ measured/approximate value	Measured value	Approximate value	Δ measured/approximate value
30	5.8	5.41	−0.39	6.2	5.79	−0.41	6.4	6.17	−0.23
40	6.9	6.81	−0.09	7.1	7.3	0.2	13.9	–	–
50	8.8	8.38	−0.42	9.3	9.02	−0.28	19.3	–	–
60	9.9	10.11	0.21	10.5	10.94	0.44	22.7	–	–
70	12.2	12	−0.2	25.8	–	–	31.3	–	–
80	14.2	14.05	−0.15	32.7	–	–	39.6	–	–

For safety reasons, we recommended to use the value of $+10\%$ when programming the calculated puncture length.

Based on the tested piercings, this value seems to be ideal and universal for all material thicknesses in the range of 15–80 mm. That means, if, e.g. for a material thickness of 15 mm and the selected jet feed rate of 5 mm·min^{-1} along the machined surface of the cut material during piercing, the calculated piercing length was 2.74 mm, during programming and performing the AWJC, the piercing length (piercing trajectory length) would be 2.74 mm $+$ 10% $=$ 3 mm.

4.4 Experimental Cutting by Using the Designed Method of Forming a Piercing on the Workpiece Contour

After determining the mathematical formula, it was possible to experimentally verify the formation of a piercing on the workpiece contour. Thicknesses of the DIN 1.4301 material selected for cutting were in the range of 15–80 mm.

4.4.1 Designing a Subprogram Within a Production NC Program for Forming a Piercing on a Workpiece Contour, and Experimental Cutting

To make a piercing on the workpiece contour, it is necessary to enter the piercing trajectory length (piercing length) into the production NC program. The programmer will design a program with subprograms containing (in digital form) the trajectory lengths determined by the designed equation, along with the jet movement velocities for individual DIN 1.4301 material thicknesses. For practical reasons, the piercing lengths in the programs are numerically rounded upwards, to one decimal place. Programmer then saves the developed subprograms for their further use.

An NC entry of a piercing formation on the workpiece contour:

The subject of cutting is a certain shape of DIN 1.4301 material and its specific thickness. Programmer determines the piercing location pre-determined in the subprogram. The measured values of the point coordinates are then entered in the beginning of the program. After that, particular pre-designed subprogram is added to the program with respect to the cut material thickness. Finally, the coordinates of starting the cutting of a specified shape are supplied in the end of a subprogram, and entered is a particular program of the subsequent cutting the contour shape of the future workpiece. The NC program was modified to perform the piercing on the workpiece contour, together with the specified piercing length. We can state that the completed cutting met the established goals. Compared to the standard cutting method and the approach methods used (perpendicular, radius, angular piercing, etc.), the piercings were formed on the workpiece contour (Fig. 4.5), thereby eliminating the formation of residual material (Fig. 4.6).

Subsequently, the cutting of the annulus (Fig. 4.7) was tested under the same selected technological conditions of cutting. Cutting the annulus with a piercing formed on the workpiece contour is important, inter alia, in particular for material

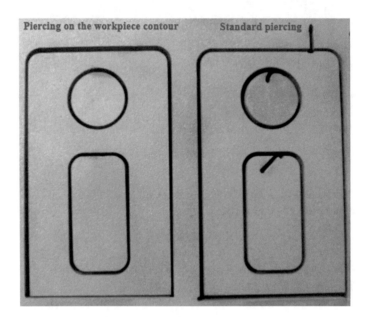

Fig. 4.5 The experimentally cut component with standard piercings

Fig. 4.6 The experimentally cut component with piercing on the workpiece contour

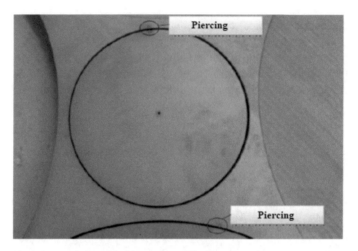

Fig. 4.7 Annulus—a detail view of another performed piercing

savings, since no residual material is formed when piercing, and the saved material can be then used to cut other parts.

4.4.2 Piercing the DIN 1.4301 Material of 15–80 mm Thickness Range

Within the experiment, other thicknesses of the DIN 1.4301 material in the range of 15–80 mm were cut. The objective was to make a piercing on the contour workpiece of the cut material, while assessing the following parameters:

- Formation of a piercing on the workpiece contour,
- Extent of damage to the workpiece shape and residual material,
- Quality of the machined surface in the area of the made piercing.

The cutting was performed under the technological constant parameters (Chap. 4.3) and variable parameters (Table 4.6).

The jet feed rate along the machined surface of the cut material during piercing and the working speed were chosen with respect to the thickness of the cut workpiece. The dimensions of the cut workpieces were chosen arbitrarily. The piercing length calculated using Formula was +10%.

The magnification scale of the piercing area details (in Figs. 4.8, 4.9, 4.10, 4.11, 4.12, 4.13, 4.14, 4.15, 4.16) below was chosen as necessary, and it varied for each image.

	Material thickness [mm]	v_f by piercing [mm·min^{-1}]	v_f by cutting [mm·min^{-1}]	L_p [mm]
Table 4.6 Variable parameter for cutting	15	30	210	4.6
	20	10	160	3.9
	25	15	120	4.9
	30	20	90	6.0
	40	25	65	8.0
	50	20	55	9.2
	60	15	40	10.2

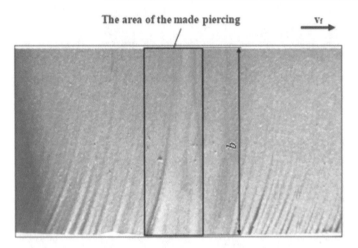

Fig. 4.8 A magnified detail of the performed piercing on the contour of workpiece 15 mm thick

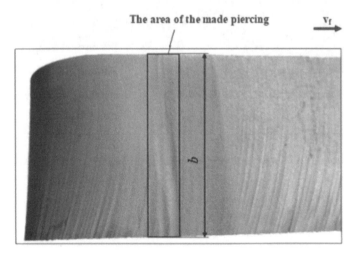

Fig. 4.9 A magnified detail of the performed piercing on the contour of workpiece 20 mm thick

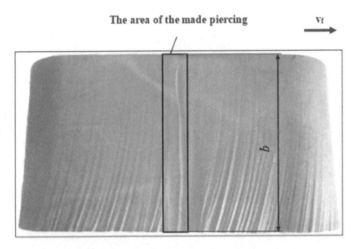

Fig. 4.10 A magnified detail of the performed piercing on the contour of workpiece 25 mm thick

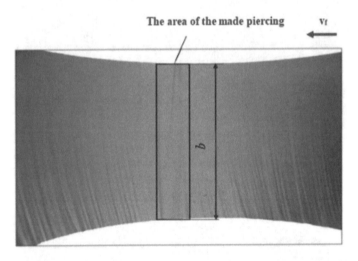

Fig. 4.11 A magnified detail of the performed piercing on the contour of workpiece 30 mm thick

Evaluation of the made piercings:

Not exceeding the deviations typical for material cutting by jet technology (STN EN ISO 9013:2017–08).

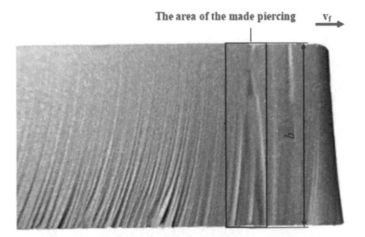

Fig. 4.12   A magnified detail of the performed piercing on the contour of workpiece 40 mm thick

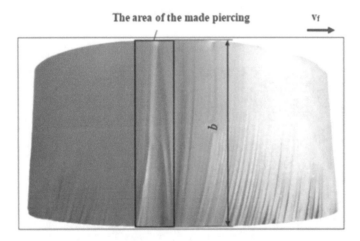

Fig. 4.13   A magnified detail of the performed piercing on the contour of workpiece 50 mm thick

4.5   Evaluation of the Material Savings When Piercing on the Workpiece Contour

Placing the piercing point directly on the future workpiece contour makes the preconditions for

- Reduced consumption of a semi-finished product intended for a particular component manufacture,
- Increased utilization rate of residual material.

Figures 4.17, 4.18 compare the material consumption

The area of the made piercing

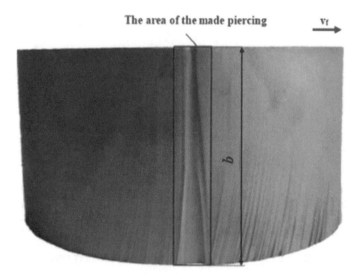

Fig. 4.14   A magnified detail of the performed piercing on the contour of workpiece 60 mm thick

The area of the made piercing

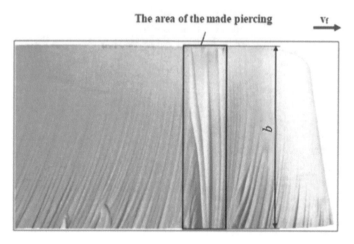

Fig. 4.15   A magnified detail of the performed piercing on the contour of workpiece 70 mm thick

(a)   when locating piercing outside the workpiece space,
(b)   when locating piercing on the workpiece contour.

   The resulting total weight of the material in b) was reduced by 12%.
   Table 4.7 lists the possibilities of the DIN 1.4301 material savings versus the size of the cut thickness of material.

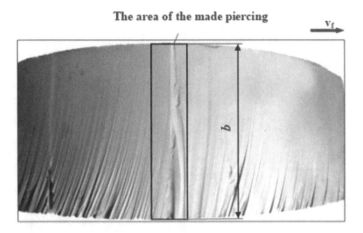

Fig. 4.16 A magnified detail of the performed piercing on the contour of workpiece 80 mm thick

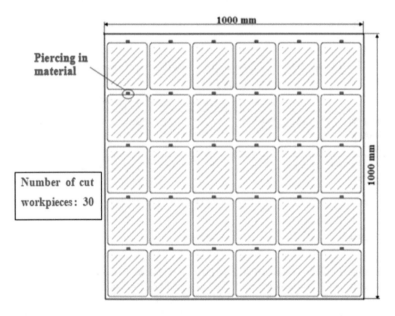

Fig. 4.17 Material consumption when locating piercing outside the workpiece space

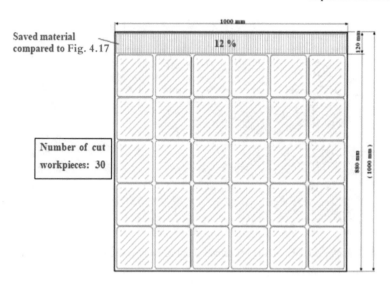

Fig. 4.18 Material consumption when locating the piercing on the workpiece contour

Table. 4.7 Savings of the DIN 1.4301 material versus the size of the cut thickness of material

	Material thickness DIN 1.4301							
	15 mm	20 mm	30 mm	40 mm	50 mm	60 mm	70 mm	80 mm
Total mass of blank 1000 × 1000 mm	120 kg	160 kg	240 kg	320 kg	400 kg	480 kg	560 kg	640 kg
Mass of the saved material 120 × 1000 mm	14,4 kg	19 kg	29 kg	38,5 kg	48 kg	58 kg	67 kg	77 kg

References

Ohlsson, L., et al. (1992). Optimisation of the piercing or drilling mechanism of abrasive water jets. In A. Lichtarowicz (Ed.), *Jet cutting technology. Fluid mechanics and its applications*, vol 13. Springer. https://doi.org/10.1007/978-94-011-2678-6_24

Hloch, S., & Valíček, J. (2008). *Vplyv faktorov na topografiu povrchov vytvorených hydroabrazívnym delením* (p. 125). FVT TU. ISBN: 978–80–553–0091–7.

Conclusion

Based on the results presented in the experimental part, the hypothesis established in the current monograph was verified and confirmed: the AWJC piercing in the material can be performed on the workpiece contour in the entire thickness of DIN 1.4301 within the 15–80 mm thickness range (cutting the larger or smaller thicknesses will be a subject of further experimental research).

When cutting workpieces with piercing on their contours, it is necessary to take into account the extra time for writing and designing the necessary NC program. However, an experienced programmer can continuously develop a library of subprograms to program piercing, thereby shortening the time of designing an NC program for repetitive or future production.

Making a piercing requires an extra work of setting the technological parameters, while focusing primarily on performing a piercing without damaging the shape of the workpiece and residual material, and achieving sufficient surface quality of piercing.

With suitably established technological parameters, the process of piercing on contour does not have a negative effect on the attained surface quality of the material surfaces pierced within the entire material thickness.

The basic principle of AWJC makes it clear that the jet deflects as the depth of material increases. Deflection of the water jet induces the occurrence of scratches in the machined material. Although, at first glance, the grooves formed in the area of material piercing in the entire thickness of the material seem to be deep, the control measurement (for each cut material thickness) revealed that the resulting surface unevenness was still within the permitted tolerance and the required accuracy class.

A piercing performed on the workpiece contour does not produce residual material. Increased intact area of the residual material increases the possibility of its further use.

We can state that the results of the performed experiments confirmed the correctness of the proposed innovative performance of piercing on the workpiece contour.

I. Kleinedlerová and P. Kleinedler, *Piercing of Materials with Abrasive Water Jet*,
SpringerBriefs in Applied Sciences and Technology,
https://doi.org/10.1007/978-3-030-92130-9

Bibliography

Anwar, S. et al. (2019). Bibliometric analysis of abrasive water jet machining research. *Journal of King Saud University - Engineering Sciences, 31*(3), 262–270.

Kleinedlerová, I., Janáč, A., & Kleinedler, P. (2011). The impact analysis of the selected abrasive water nozzle parameters shape in relation to the machined surface geometrical accuracy. In *TEAM 2011: Proceedings of the 3rd international scientific and expert conference with simultaneously organised 17th international scientific conference CO-MAT-TECH 2011, Trnava* (pp. 212–215). ISBN 978-953-55970-4-9.

Kumar, R. et al. (2016). Surface integrity analysis of abrasive water jet-cut surfaces of friction stir welded joints. *International Journal of Advanced Manufacturing Technology, 88,* 1687–1701.

Norma (Standard) STN EN ISO 9013:2017–08 (2017). *Tepelné rezanie kyslíkom. Klasifikácia tepelných rezov. Geometrická špecifikácia výrobku a tolerancie kvality. (Thermal cutting. Classification of thermal cuts. Geometrical product specification and quality tolerances),* (p. 14).

Norma STN ISO 4287–2 (014450) (1993). *Drsnosť povrchu. Terminológia. (Surface roughness. Terminology)* (p. 12).

Norma VDI 2906 (1994). *Schnittflächenqualität beim Schneiden, Beschneiden und Lochen von Werkstücken aus Metall.* Teil 2 – Abrasiv-Wasserstrahlschneiden (p. 4). Verein Deutscher Ingenieure.

Radvanská, A. (2010). Abrasive waterjet cutting technology risk assessment by means of failure modes and effects analysis method. *Technical Gazette, 17*(1), 121–128. ISSN 1330–3651.

Summers, D.A., & Blaine, J. G. (1997). *A fundamental tests for parameter evaluation.* Geomechanics 93 (pp. 321–325). Balkema.

© The Author(s), under exclusive license to Springer Nature Switzerland AG 2022 49
I. Kleinedlerová and P. Kleinedler, *Piercing of Materials with Abrasive Water Jet,*
SpringerBriefs in Applied Sciences and Technology,
https://doi.org/10.1007/978-3-030-92130-9

Printed in the United States
by Baker & Taylor Publisher Services